Entrepreneurship in Chemical Engineering 2022 and Chemical Engineering and the Law Forum 2022

Topical Conference at the 2022 AIChE Annual Meeting

Phoenix, Arizona, USA
13-18 November

ISBN: 978-1-7138-7903-9

Printed from e-media with permission by:

Curran Associates, Inc.
57 Morehouse Lane
Red Hook, NY 12571

Some format issues inherent in the e-media version may also appear in this print version.

Printed with permission by Curran Associates, Inc. (2023)

For permission requests, please contact AIChE
at the address below.

AIChE
120 Wall Street, FL 23
New York, NY 10005-4020

Phone: (800) 242-4363
Fax: (203) 775-5177

www.aiche.org

Additional copies of this publication are available from:

Curran Associates, Inc.
57 Morehouse Lane
Red Hook, NY 12571 USA
Phone: 845-758-0400
Fax: 845-758-2633
Email: curran@proceedings.com
Web: www.proceedings.com

TABLE OF CONTENTS

102a Intellectual Property and the Circular Economy

Paul Townsend, Lauren Dowty, Charles Collins-Chase

In view of the shift towards a more sustainable economy, industries of all kind are putting aside the "take-make-waste" economic model and developing circular economy technologies. The goal of the circular economy is to eliminate waste through new or improved designs of materials, products, and systems. With this renewed focus on the circular economy, companies have begun to develop a variety of new proprietary technologies that are now being brought to the market. With the increase in new circular economy technologies, the necessity for intellectual property continues to rise. Although companies need to protect their IP rights, there is also a need to allow other companies to help the world become more sustainable. Thus, companies must balance the need to protect IP rights and the need to share their technologies.

This presentation will provide an overview of the basics of IP, how IP is at the forefront of circular economy technologies, and important considerations for members of industry, academics, and anyone with an idea for a circular economy invention. Speakers will also highlight several case studies addressing sustainable IP and whether the IP was strong enough.

102b Protecting Ideas in Joint Research Projects

Robert Sovesky

The discussion will include a general overview of the different types of joint research arrangements, such as, Joint Development Agreements (JDAs) and Cooperative Research andamp; Development Agreements (CRADAs). The discussion will particularly focus on considerations prior to entering into the joint research agreement, considerations during the joint research agreement, considerations after termination of the joint research agreement, and intellectual property (patents, trade secrets, etc.) that may be generated throughout the process. The discussion will include practical tips for protecting ideas for individual inventors, emerging growth companies, and large corporations.

174a Innovation Fellowship: Novel Postdoctoral Training to Build a Pipeline of Innovation-Minded Professionals and Promote Commercialization in Academic Units

Anastasiia Visheratina, Dylan Neale, Kaylee Steen, Elizabeth Herness Peters, Nadine Wong, Joerg Lahann

The Biointerfaces Institute (BI) at the University of Michigan (U-M) is rooted in the mission to advance biomedical discoveries by disrupting the status quo of isolated science and fostering cross-disciplinary technological breakthroughs in nanotechnology, advanced materials, cell engineering, and neural engineering. By co-locating 28 research groups from twelve departments across the U-M schools of Engineering, Dentistry, Medicine, and Pharmacy, forward-thinking collaborations lead to healthcare breakthroughs and advanced learning. To further accelerate the path from basic research to real-world health outcomes, the BI has created a novel postdoctoral fellowship to train the next generation of scientists in commercialization and translation. The Innovation Fellowship provides a blueprint for biomedical research units to train and mentor individuals in a systematic manner that encourages out-of-the-box thinking, experiential learning that supports commercialization efforts, and enhances the network of innovation leaders.

The Innovation Fellowship focuses on five main objectives, with the help of its strategic partners, to develop the Fellows' professional competencies in translation and commercialization. Fellows will receive (1) Innovation andamp; Entrepreneurship Training, (2) Networking and Mentorship, (3) Professional Development Opportunities, and engage in (4) Technology Assessment and (5) Facilitate Project Development at all stages of the commercialization pipeline. In January 2022, BI launched the first of its kind Innovation Fellowship with Dr. Dylan Neale and Dr. Anastasia Visheratina as the inaugural cohort Fellows. During the term, fellows are trained to become a champion of innovation and translation at BI, advance BI's commercialization resources, and collaborate with BI research groups to uncover new projects with commercialization potential. They also engage with stakeholders to understand the opportunities and challenges of potential innovations as well as work with BI teams to advance projects.

Collectively, the Innovation Fellowship is creating a new pipeline of scientifically trained individuals and developing them as commercialization thought leaders to advance precision medicine, detect diseases earlier, restore body function, and improve drug delivery. In my talk, I will give a thorough overview of the Innovation Fellowship, our results, and its influence on my career development.

AUTHOR INDEX

NOTES

NOTES

NOTES

NOTES

NOTES

NOTES

NOTES

NOTES